José Ferrer
Raquel Alejos

Science and technology of cashew agro-industrial processing

José Ferrer
Raquel Alejos

Science and technology of cashew agro-industrial processing

Manufacture of agro-industrial products: pasteurized juice, nectar, liqueur, jelly and jam from cashew pseudo-fruit

ScienciaScripts

Imprint
Any brand names and product names mentioned in this book are subject to trademark, brand or patent protection and are trademarks or registered trademarks of their respective holders. The use of brand names, product names, common names, trade names, product descriptions etc. even without a particular marking in this work is in no way to be construed to mean that such names may be regarded as unrestricted in respect of trademark and brand protection legislation and could thus be used by anyone.

Cover image: www.ingimage.com

This book is a translation from the original published under ISBN 978-620-6-76164-8.

Publisher:
Sciencia Scripts
is a trademark of
Dodo Books Indian Ocean Ltd. and OmniScriptum S.R.L publishing group

120 High Road, East Finchley, London, N2 9ED, United Kingdom
Str. Armeneasca 28/1, office 1, Chisinau MD-2012, Republic of Moldova, Europe
Printed at: see last page
ISBN: 978-620-8-03021-6

SCIENCE AND TECHNOLOGY OF CASHEW (ANACARDIUM OCCIDENTALE L.) AGRO-INDUSTRIAL PROCESSING

JOSÉ RAMÓN FERRER GONZÁLEZ
RAQUEL ELENA ALEJOS PINEDA

SUMMARY

SUMMARY

The physicochemical characteristics of the pseudofruit and juice of yellow and red cashews (Anacardium Occidentale L.) from a plantation located in the Zulia Fruit Centre, Venezuela (11° 00' N, 71° 30' O) were evaluated. A completely randomised factorial arrangement was used in a split plot design (22 x 4 x 10), with hierarchical sampling effects at harvest time. An analysis of variance (PROC- GLM) and a separation of means (LS MEANS procedure) were carried out using SAS. The yellow pseudofruits without the nut had a higher average mass ($p<0.05$) compared to the red ones. Differences were detected ($p<0.05$) in the °Brix of the juice of pseudofruits from selected plants, probably due to genetic variability. Pseudofruits harvested from February to March showed higher average values ($p<0.05$) in all variables except firmness, pH and maturity index. The differences observed during sampling ($p<0.05$) in the mass of the pseudofruits and in the maturity indices were due to the climatic conditions in the area at all harvest times. Firmness, maturity levels and titratable acidity were affected by the Type x Period interaction, while the Type x Sampling interaction affected all variables except mass with and without nuts and maturity level. The high mass yield of the seudo fruit and the physico-chemical characteristics of the juice guarantee a good supply of high-quality raw material for the industrialisation of this tropical fruit.

Key words: Anacardium Occidentale L., physico-chemical characteristics, quality.

INTRODUCTION

The major challenges faced as a result of the globalisation of the economy become development opportunities for countries as they make efficient use of resources to generate benefits.Cashew (Anacardium occidentale L.) is a fruit of great economic importance due to its nut, which represents only 10% of the fruit. The rest, known as the pseudofruit, is discarded [1,2]. The name cashew (Anacardium) refers to the shape of the fruit, which resembles an inverted heart (ana means "up" and cardium "heart") [3]. The pseudo-fruit has been classified by several researchers as a food with high nutritional value, due to its high nutrient content, which makes it a good choice as a raw material for the production of foods with high nutritional value and a palatable flavour for consumers [4].

Due to Venezuela's agro-ecological characteristics, national fruit production has expanded significantly in recent years, mainly due to the restrictions imposed on the process of importing fresh and semi-processed fruit, the increase in demand and supply of good quality fruit and the development of processing industries for traditional tropical fruits such as citrus fruits, pineapples and bananas [5].

However, this economic activity is one of the productive sub-sectors that has developed in the country. There is therefore a need to define and implement policies that take into account the interests of the marketing chain in this agricultural sector, in order to position our tropical fruit on foreign markets and, above all, in industrialised countries, improving the technological level of existing plantations and diversifying fruit production [6].

Cashew cultivation, a species that has been considered within the group of "minor fruit trees", is presented as one of the most promising alternatives for expanding and strengthening the development of the national fruit sector [7].

This fruit tree, native to tropical America, is agronomically undemanding and can be fully exploited. A wide variety of products with high nutritional and commercial value can be obtained from its industrial use, which generates a source of foreign currency for the economy [8].

In Venezuela, the development of plantations has been promoted in some eastern states, specifically in southern Anzoátegui, northern Bolívar, Monagas and part of Guárico, where around 17,000 ha of this crop are registered, while in the north-western area of the state of Zulia there are approximately 900 ha of Creole-type cultivation, which is emerging as an option with great potential, despite unflattering previous experiences [9].

In order to promote the technical development of cashew plantations and their agro-industrial use, the physico-chemical characterisation of cashew pseudofruit and juice was carried out, using two types of cashew (red and yellow), to assess their agro-industrial potential and develop the technological blueprint for the industrial production of pasteurised juice, nectar, liqueur, jelly and jam.

CHAPTER I
REVIEW BIBLIOGRAPHY

1.1 General

Cashew (Anacardium occidentale L.) is native to the north-east of Brazil, in the Maranon region. It was taken by the Portuguese to their colonies in Asia and Africa, and the Spanish introduced it to Central America [9].
This species is a plant that can be grown in geographical areas with low soil fertility and low water demand. It is a typical tree of tropical climates, found from Mexico to Peru, including Hawaii, Puerto Rico and parts of southern Florida [9].
The cashew tree is mainly valued for its nuts, which are eaten roasted and salted and sold in vacuum packs. Cashew nuts are considered a luxury product compared to peanuts, pistachios and almonds.It is recognised worldwide as the most commercially valuable nut due to its pleasant taste, high nutritional value, high calorie content, rich in proteins, fats, phosphorus and thiamine, and minimal amounts of cholesterol. The nut trade is dominated by two countries, Brazil and India, which are the biggest exporters. So far, the biggest buyer has been the United States. On the international market, the price of almonds is around US$ 3.00 per pound for whole almonds, US$ 2.50 per pound for halves and US$ 1.50 per pound for pieces of almonds [10].
It is important to note that in Venezuela it grows spontaneously and wild in different regions, with important plantations in Bolívar State, Anzoátegui State and Monagas State. Production from these plantations was transferred to the Complejo Mereyero Pariaguán in Anzoátegui State, a factory designed with advanced technologies in which only the nut (the fruit of the plant) was processed in order to obtain roasted nuts and their oil, which was exported to foreign markets. In the process, the pseudo-fruit became a waste material [6].
In cashew nut industrial processing centres, approximately 90% of the pseudo-fruit remains as residue or waste, and only 5% of it is used by some artisan micro-enterprises to obtain products such as "merey pasado", made by osmotic dehydration of the pseudo-fruit; soft drinks and jellies from the juice; liqueur, vinegar, from the alcoholic fermentation of the juice and marzipan from the ground nut [11].The consumption of ripe cashews is recommended both for their flavour and medicinal properties and for their high nutritional value, especially in relation to their high vitamin C content [12], as well as their minerals and other vitamins [13].

1.2 agronomic aspects

1.2.1 Taxonomy

Class: Angiospermae. Subclass: Dicotyledoneae. Order: Sapindae.
Family: Anacardiaceae.

Kingdom: Vegetable. Genus: Anacardium. Species: Occidentale L.
Scientific name: Anacardium occidentale L.

Common names: Merey, caujil (Venezuela), Cashew (English), Acajú (French) and Cajú, castaña de cajú (Portuguese).

There are approximately 390,000 plant species in the world, and Brazil has one of the greatest biological diversities in the world, with more than 46,000 plant species. This Biodiversity is still little known and its use as food has been neglected. Its use in the local diet contributes to broadening the sources of nutrients available to the population and to promoting food sovereignty and security [14].

Cashew is grown especially in Benin, Brazil, Guinea-Bissau, India, Indonesia, Ivory Coast, Nigeria, the Philippines and Vietnam [15]. The nut is the real fruit of the cashew, consisting of the pericarp (shell) and the kernel. In addition, the peduncle, also known as the cashew apple, is a pseudofruit and accounts for 90 per cent of the fruit's weight. The cashew pseudofruit is considered a by-product of the cashew industry, since the nut represents 10% of the cashew industry's weight and is the industry's raw material [15].

Cashew (Anacardium occidentale L.) is one of the most useful tropical tree species and is a very important alternative for the Colombian Altillanura, where there are comparative advantages for large-scale planting of this species [1].

Widely distributed and often cultivated in Venezuela, it is nevertheless a fruit tree that has not been technologically exploited. Carvajal, in 1648, mentions it in Apure and describes that its fruits are the size of an egg and have a bittersweet flavour; Caulin then mentions in his history that "they are cultivated in many parts of these provinces of Cumaná, Guayana and on the island of Trinidad under the name of Merey". Their plantations are characterised by trees that grow rapidly and can reach 6 to 10 metres, are always green and are easily recognised by their bright and colourful foliage [16].

It is considered a melliferous plant, due to the attraction of its flowers for bees, and has also been used to shade some plantations and as an ornamental tree [6].

1.3 Characteristics agroclimatic

1.3.1 Conditions weather

Cashew plants adapt easily to highland soils, with altitudes between 50 and 800 metres above sea level. At an altitude of more than 600 metres above sea level, there is a marked decrease in production, while at an altitude between sea level and 400 metres above sea level, higher production is observed [6].
It grows with rainfall ranging from 500 mm to 3,800 mm, spread over a period of 5 to 7 months, since a well-defined and sufficiently long dry season allows flowering and fruiting, while rainfall of less than 800 mm results in irregular fruiting [8].
The minimum relative humidity favourable to its development has been observed to be 46-56% and a maximum relative humidity of 68-78%. Its trees grow favourably at a temperature between 18 °C and 30 °C; it has been observed that when the plantation is young, it is very susceptible to cold and the average temperature is 26 °C [17].

1.3.2 Soil

Cashew cultivation is one of the best alternatives for agrosocioeconomic development in areas with unfavourable soil and climate conditions, contributing to reforestation and soil conservation, with certain ecological benefits in the future. Although it is not a demanding crop in terms of soil fertility, its preference for light, sandy, deep and well-drained soils has been proven, as it is very sensitive to flooding.It can grow in stony and semi-arid soils; it is abundant in savannas and tolerates drought. Sandy-clay soils with plenty of organic matter and great depth are recommended, effective in preventing uprooting in windy areas; it is less tolerant of saline soils than most coastal plants [18].
With regard to soil pH, Northwood points out that it adapts to a wide range, but that in slightly acidic soils its development and production are greater. Its growth can be observed in soils with pH values between 4.3 and 8.7 [19]. In general, it is not necessary to set up an irrigation system; however, in situations where annual rainfall is less than 1,000 mm, surface irrigation should be used [6].

1.3.3 Flowering

Flowering is observed from December to February; however, there are reports that the flowering period has been recorded from February to May. The fruiting and ripening period can take place from April to June and may extend into October. However, harvests were recorded from January to March. As fruiting depends on climatic variations, it can be concluded that the fruit ripens 2 months or 2 and a half months after flowering [6].

1.3.4 Fruits

The fruit has a very peculiar shape, consisting of a kidney-shaped, glossy brown nut (edible when roasted), 2 to 4 cm long, on a thick pear-shaped stem or false fruit, bright yellow or scarlet red, juicy, edible, with a sour, slightly astringent flavour, 6 to 10 cm long by 4 to 6 cm wide [6].

1.3.5 Fruiting

Cashew trees start bearing fruit after two or three years of age and reach full production between five and seven years of age. It was possible to demonstrate that a plantation, where all the necessary cultural practices are carried out for its development, can start fruiting (on a very small scale) from the first year, hoping to obtain its full production. higher production before the age of five. These results were observed and obtained in the plantation being developed by the Centro Frutícola del Estado Zulia, owned by CORPOZULIA, Venezuela [6].

1.3.6 Crop varieties and production areas

Countries like Brazil have extensive knowledge of the crop, both from an agronomic point of view and in terms of physical and chemical quality, which is why, through genetic improvement, they have obtained clones within the different types of cashew nuts that are grown in this region [20]. This has allowed them to classify them according to the size of the nut, making it the second largest exporter of cashew nuts, which generates a source of foreign currency for the country's economy. Similarly, the processing of the cashew pseudo-fruit to extract juice and make soft drinks, liqueurs and other products with high yields has been technologised [21].

Brazilian research into this crop has led to progress in the genetic improvement of the early dwarf cashew tree at the Pacajus Experimental Station. These plants are characterised by their small size, high production and require improved management techniques [22].

In Venezuela, there are no specific varieties as a result of seed propagation, but only two defined types, the scarlet red cashew and the yellow cashew. It is rarely cultivated, but is found in semi-wild form. Similarly, there are currently no improved varieties of the common type of cashew [6].

In Colombia, the existing varieties are the product of natural crossbreeding. In general terms, the colour and size of the flesh, the fleshy part or "apple", are very different. The main varieties are: the giant Magdalena, yellow and red; the long Nazaré and the small Meta [6].

In Nicaragua, two types of fruit are currently known, red and yellow apples, the latter being less astringent than the red ones. Table 1.1 shows the areas of cashew cultivation in the world [6].

Table 1.1

International cashew growing areas (Anacardium occidentale L.)

Continent	Geographical location
Americas	United States (Florida), Mexico, Cuba, Haiti, Jamaica, Guatemala, Antilles, Salvador, Trinidad, Venezuela, Colombia, Peru, Brazil, Panama.
Africa	Senegal, Mali, Guinea, Ivory Coast, Ghana, Dahomey, Nigeria,Kenya, Congo, Tanzania, Angola, Mozambique, Madagascar, South Africa.
Asia	India, Ceylon, Indochina; Philippines, Malaysia, Indonesia.
Oceania	Hawaii, Tahiti, Australia

1.3.7 Harvesting

The nuts are harvested by hand when they change colour from green to grey and the pseudo-fruit flower has developed and is close to fully ripening. During harvesting, the fruit must be treated with care, as bruising causes them to lose their commercial value when they are marketed as fruit for consumption, either fresh or for industrialisation [6].

1.3.8 Post-harvest

Post-harvest, cashews are the source of various products, both primary and secondary. Primary products are those obtained on the plantation, such as the pseudo-fruit and the nut; secondary products or by-products are those in which the primary products constitute the raw material for certain industries, responsible for the manufacture of commercial items, such as the pulp of the pseudo-fruit, the nut kernel, the "cardol" oil from the nut shell and others [6].

1.3.9 Production

In Brazil, the leading producer of the fruit, two types of cashew are known: the "common", with a large tree that produces a large nut and a medium yield (medium kernel-to-nut ratio), and the "dwarf", with a smaller tree than the "common" but a high yield (high kernel-to-nut ratio). The "dwarf" cashew is also earlier than the "common", as it starts production a year earlier than the "common" (2 years) [21].

The yield of a three-year-old ordinary cashew tree is 3.17kg of nuts per tree. At 15 years of age, it will have a yield of 31.7kg of nuts per tree. Mature trees of exceptional quality have a yield of 90.71kg of nuts per tree. Many trees produce nuts from 15 to 20 years of age, others are known to be productive up to 45 years of age, without forgetting that, by mass, cashew is made up of 10% nuts and 90% pseudofruit [21]. In Venezuela, there is no precise data on the production of this fruit tree [6].

1.4 Use agroindustrial

Cashew cultivation is perhaps one of the most useful plants in existence, as it can be used almost in its entirety [6]. It is mainly appreciated for its seeds or nuts (true fruit), which are commercialised all over the world. Its kernels have a high protein content and its oil has many industrial applications. Cashew is the nut with the lowest fat content (43 per cent) compared to the eight best nuts on the international market, and shares first place with pistachio in terms of fat content. protein (21%). It represents approximately one third of the nut's weight and its analysis indicates an oil content of 55 to 60 per cent, 15 to 20 per cent protein and 5 per cent carbohydrates (starch and sugar). They are eaten roasted and salted and are sold as "pasapalos". A type of nougat is prepared from these

kernels, which is very pleasant [6].Walnut shell liquid (LCN) is obtained from walnut shells, which has many industrial uses in the manufacture of cosmetics, resins, varnishes, colourants, dyes, electrical insulation material, brake pads, adhesives and many others, as well as being used in confectionery and baking. In the first countries to buy this product (the United States, England and Japan), there are 243 patents for products in which LCN appears in the formulation [6].
The husk of the seed has a very oily sap, i.e. a caustic oil, which evaporates during roasting. Its active ingredient is anacardic acid or cardol oil, dark brown in colour and pungent in taste, which is highly caustic and poisonous. It is used in industry as a medicine to protect carved wood, book covers and various articles from insect attack [6].
Its pseudo-fruits (stalks) are used for fresh consumption and are rich in vitamins. There are various products obtained from its processing. However, it is estimated that the industrial use of cashew pseudofruit is very low (approximately 5%), which translates into a significant loss of its industrialisation potential. These losses are mainly due to the fact that the pseudofruit has the particularity of being highly perishable (it ripens before the nut), it has a small production capacity for industry, its harvest period is short and there are no economical methods for its conservation. It has been established that these losses can be considerably reduced (92.6% to 78.5%), depending on the number of plants in the plantation and the type of variety or selection [6].
The pomace that remains after extracting the juice is used to feed livestock or can be processed into flour for human and animal consumption. The protein quality of cashew meal is very high and it can be included without restriction in pig feed; however, it can cause diarrhoea in calves and dairy cows if given in large quantities [6].
The wood is used in some countries, on a small scale, to produce charcoal, in construction, carpentry, boats and more. The cashew tree is also used as an ornamental tree in gardens. A gum similar to "arabica" is extracted from the trunk, which is used to bind texts and also serves as a moth repellent. An indelible ink is prepared from the milky sap [6].
In Venezuela, the cashew pseudo-fruit is used in the production of "merey pasado" and syrup, products that have the potential to be industrialised for consumption not only nationally, but also abroad, through effective promotional campaigns [6].

1.4.1Processing artisanal

Although cashews have a high food value, their percentage of industrialisation is very low. However, in Brazil, this fruit is used to make a variety of products, which can be handmade or made by small and medium-sized companies, such as mass-produced sweets, sweets in syrup, jam [23], frozen pulp, cajuína, gels and syrup, as well as cashew juice, a delicious and refreshing drink that is attributed medicinal qualities and whose vitamin C content is four times higher than that of oranges; it is estimated that a glass of cashew juice meets all the daily vitamin C needs of an adult person. In Venezuela, it is eaten fresh or in sweets. In some regions of Western India and tropical America, where cashews are grown only occasionally, it is customary to use the pseudo-fruit and discard the nut [24].

1.4.2Processing industrial

The main cashew pseudofruit products produced on an industrial scale are: whole juice with a high pulp content, concentrated juices, cajuína, nectar, preserved pulp, frozen pulp and sweets, with the greatest potential for industrial use in products such as cashew wine, dried cashews, dehydrated cashews [25], clarified and carbonated juice, chutney. The juice has a high pectin content and can be used to make preserves and jellies, as well as containing a considerable amount of sugars, tannins and minerals [26]. A large number of products can be obtained from the small-scale industrial processing of this fruit, including jams, jellies and a wide variety of dishes and desserts, as well as spirits and fermented drinks [21].

The total sugar content in cashew pseudofruit is of great importance for the production of liqueur from its juice. The alcohol yield and therefore the type of wine obtained (high or low alcohol) is defined according to the sugar content, which will be converted by alcoholic fermentation by yeasts into alcohol and carbon dioxide. The production of alcohol by this biochemical route is around 50 per cent of the fermented sugars. Must wine is a drink obtained by fermenting the juice of the must, which is then clarified, filtered and presented clean and bright. In this drink, the alcohol content can vary from 10 °GL to 14 °GL, obtained by fermenting the clarified and corrected juice of fresh, healthy sweet cashews. This product is not currently manufactured on an industrial scale, but the Ceará Industrial Technology Centre, in conjunction with EMBRAPA Agroindústria Tropical and the Federal University of Ceará, Brazil, is carrying out research into obtaining five types of cashew wine from the

NUTEC pilot plant, in post-graduate work conducted by EMBRAPA technicians. Alcohol can be obtained from the distillation of the product obtained from the alcoholic fermentation of cashew juice. To obtain brandy, the distillation goes through a maturation and ageing process [21]. Cashew derivatives and industrialised whole juice are of great economic importance in Brazil and are widely accepted on the domestic market. Nectar, ready-to-drink juice, is also found on the Brazilian market, presented in long-life packaging, with great acceptance, mainly due to its ease of consumption [6]. In the Venezuelan regions of Soledad and Ciudad Bolivar, the pseudo-fruits of the cashew are used to make jam, dried in a hot air oven at approximately 20% humidity, intensifying their flavour and improving their stability for consumption as sultanas, and the famous "Dulce de Merey en almíbar" and "Merey pasado" are made, which are served on special occasions [6].

1.5 Agro-industrial pilot plant from food

This is a small-scale facility designed to carry out experimental tests and develop food production processes at an industrial level. These plants allow concepts to be tested, processes to be adjusted and the technical and economic viability of new technologies or food products to be assessed before they are implemented on a large scale in industry.An agro-industrial pilot plant is designed by developing conceptual engineering, basic engineering procedures and finally detailed engineering to carry out food production. For the agro-industrial processing of food products, the requirements established by Good Manufacturing Practices (GMP) and the Codex Alimentarius are taken as a basis. [40, 41].Good Manufacturing Practices (GMP) establish the requirements and fundamentals needed to guarantee safety at all stages of the food production process. They must consider all aspects covered by GMP, such as infrastructure (physical facilities), hygiene measures, equipment and utensils,personnel, raw materials, operations and the process cleaning system at the design stage.

1.5.1Facilities physical

The basic design characteristics that must be met by the physical facilities of a food processing plant are established by the Codex Alimentarius, specifically in standard RTCA 67.01.33:06 [40]. They are described below:

•Factory buildings and structures must be of a size, construction and design that facilitates maintenance and sanitary operations to fulfil the purpose of food processing and handling, protection of the finished product and cross-contamination.

- Food industries must be designed so that they are protected from the outside environment by walls. Buildings and facilities must prevent the entry of animals, insects, rodents and/or pests or other environmental contaminants such as smoke, dust, vapour or other contaminants.
- The building's environments should include a specific area for changing rooms, with appropriate furniture for storing personal items, and a specific area for eating.
- The facilities must allow for easy and adequate cleaning and inspection.
- Drawings or sketches of the physical plant must be available to localise the areas related to the production process flows.
- All building materials and installations must be of such a nature that they do not transmit undesirable substances to the food. Buildings must be of solid construction and kept in good repair.
- In the production area, wood is not permitted as a building material.

As for the processing and storage areas, it is stipulated that:

1.5.2 Flooring

- Floors must be made of impermeable, washable and non-slip materials that are not toxic for their intended use and must be constructed in such a way as to facilitate cleaning and disinfection.
- Floors must not have cracks or irregularities in their surface or joints.
- Joints between floors and walls should be rounded to facilitate cleaning and prevent the accumulation of materials that could lead to contamination.
- Floors must have drains and a suitable slope to allow water to drain away quickly and prevent puddles from forming.
- As appropriate, floors should be made of materials resistant to deterioration caused by contact with chemicals and machinery.
- Warehouse floors must be made of material that can withstand the weight of the stored materials and forklift traffic.

1.5.3 The walls

•External walls can be built from concrete, brick or concrete block and even prefabricated structures made from various materials.
•Internal walls must be lined with impermeable, non-absorbent, smooth, easily washable and disinfectable materials, painted in a light colour and free from cracks.
•When justified by the humidity conditions during the process, the walls must be covered with a washable material up to a minimum height of 1.5 metres.
•The joints between walls and floors must be concave.

1.5.4 The ceilings

•Ceilings should be constructed and finished smoothly to minimise the build-up of dirt, condensation, mould and crusts that can contaminate food and the shedding of particles.
•False ceilings are permitted and must be smooth and easy to clean.
•The ceilings must be at least five metres high to prevent vapours condensing on the pipes and thus the release of dirt and dust particles accumulated in the pipes and lighting fittings.

1.5.5 Windows and doors

•Windows should be easy to clean, constructed to prevent water and pests from entering and, where appropriate, fitted with insect screens that are easy to remove and clean.
•Window openings should be sloping and of a size that prevents dust accumulation and use for storage.
•Doors should have a smooth, non-absorbent surface and be easy to clean and disinfect. They should open outwards, be well fitted to the structure and in good condition.
•Doors giving access to the outside of the processing area must be protected to prevent pests from entering.

1.5.6 Lighting

• The entire establishment must be lit by natural or artificial light in a way that allows tasks to be carried out and does not compromise food hygiene; or by a mixture of both, ensuring a minimum intensity of 540 lux (50 candelas/ft^2) at all inspection points, 220 lux (20 candelas/ft^2) in processing rooms and 110 lux (10 candelas/ft^2) in other areas of the establishment.

• Lamps and all artificial lighting fixtures located in the raw material receiving, storage, preparation and food handling areas must be protected against breakage. Lighting must not alter colours. Electrical installations, if external, must be covered by insulated pipes or tubes, and overhead cables are not permitted over food processing areas.

1.5.7 Ventilation

• Adequate ventilation must be provided to: prevent excessive heat, allow sufficient air circulation, avoid condensation of vapours and remove contaminated air from different areas.
• The direction of air flow should never be from a contaminated area to a clean area, and ventilation openings should be protected to prevent contaminants from entering.
Once the product derived from the cashew pseudofruit has been characterised at laboratory and sensory evaluation level, the next step must be taken:

o **Supply and demand study**

The supply and demand study should be developed taking into account the results obtained from the purchase intention question in the sensory evaluation survey for commercialisation. Based on this information, and given the still notable absence of this sector on Venezuelan shelves, the idea of building and implementing this plant is justified.

o **Marketing channels**

Among the marketing channels, we highlight the direct distribution of the product from the production plant to warehouses, mini-markets and supermarkets, through sellers who have their own means of transport, thus minimising the costs associated with transport and stages in the marketing chain. Social media marketing has become an efficient and high-impact advertising medium. What's more, its low cost makes using this communication tool even more attractive. With this in mind, we suggest the use of social networks such as Instagram, Twitter and Facebook as a means of mass promotion of the product; and as a marketing alternative, the use of commercial websites with a high national impact, which are free for a certain period of time. or, alternatively, very low-cost adverts for the sale of products, such as OLX and Mercado Livre, respectively.

- **Size and location**

To establish the size of the plant, a production capacity of one tonne per day (1 Ton/day) should be used as the basis for calculating the project. The location of the plant depends on certain factors, such as: location of the consumer market, sources of raw materials, availability and characteristics of labour, transport facilities and suitable communication routes, availability and cost of electricity, fuel, availability of public services (water, telephone) and a space for waste disposal [23, 25].

The plant must be located in an area where access to cashew plantations is easy. Due to the low water and care requirements that cashew plants need for their development, and considering the need for economic development of the indigenous communities originally from the state of Zulia, the northern zone of the state, the eastern coast, and part of the southern zone of Lake Maracaibo, where there are already plantations of this fruit, would be ideal areas for the development of the plant. One of the advantages of locating the industry in these areas would be the generation of sources of income for the local population. The consumer market, at first, would be the neighbouring populations, to then extend distribution to the rest of the state and even the rest of the country. With regard to the availability of labour, it would be necessary to offer some training lectures to the staff who will be working in the industry, because in order to guarantee the safety of the final product, it would be advisable to instruct the staff in the basic principles of good manufacturing practices, as well as the critical points in the process of making the product. Finally, a description of the production process, the distribution of equipment and an economic study of the agro-industry should be developed.

1.6 Physico-chemical and microbiological properties

Due to its physical and chemical properties (pH, acidity, high sugar content, humidity and low oxidation-reduction potential), cashew pseudofruit juice practically does not allow the development of fungi, yeasts and acid-tolerant bacteria. During the production process, cashew pseudofruit pulp has problems with rapid fermentation, caused by contamination by yeasts associated with it. Prolonged exposure of the pulp to the environment (oxygen and light) causes it to darken due to the degradation of vitamin C and other compounds present in the pulp. Table 1.2 shows the chemical composition of cashew pseudofruit, as presented by various authors [26, 27, 28, 29].

Table 1.2

Chemical composition of cashew pseudofruit

Composition	Quantity
Calorific intake (kJ)	125,50
Moisture (g)	88,50
Protein (g)	0,90
Total fat (g)	0,10
Carbohydrates (g)	7,70
Crude fibre (g)	2,50
Ash (g)	0,30
Calcium (mg)	14,00
Phosphorus (mg)	24,00
Iron (mg)	0,40
Sodium (mg)	16,00
Potassium (mg)	565,00
Magnesium (mg)	260,00
Zinc (mg)	5,60
Copper (mg)	2,22
Manganese (mg)	0,83
Vitamin A* (IU)	50,00
Vitamin A* (ER)	5,00
Vitamin B1 (thiamine) (mg)	0,03
Vitamin B2 (Riboflavin) (mg)	111,50
Vitamin B3 (niacin) (mg)	0,40
Vitamin C (mg)	200,00
Total extractable polyphenols (mg)	168,25

* **IU**: The International Unit used to measure the biological activity of many vitamins, hormones, enzymes and medicines.

***ER**: It's a unit called retinol equivalents.

1.6 Nutritional and medicinal aspects

The cashew (Anacardium occidentale L.) has a nut that is sold roasted, with excellent concentrations of oleic acid (Omega 9) and other phytochemicals with nutritional and antioxidant characteristics. In Venezuela, the Criollo Red and Yellow cashew is found wild and cultivated in family gardens in the east of the country and in various areas of the municipalities of Maracaibo, Mara and Rosario de Perijá, in the state of Zulia. An excellent adaptation of these creole materials to the agro-ecological characteristics of the growing areas has been reported, since the plants are in dry conditions, which has the advantage of saving money when setting up irrigation systems, a technology that is very expensive [20, 30, 31].

In the physicochemical assessment of the nut, it was determined that there is an oil extracted from the endosperm, which is bright yellow in colour and has low viscosity, containing excellent concentrations of essential fatty acids such as oleic acid (Omega 9), linoleic, stearic, elaidic and palmitic. Other important nutrients have been reported in cashew fruit, such as minerals, proteins, sugars and vitamins. In addition, some other compounds with antioxidant properties (flavonoids, total phenols, vitamin C) have been identified, which confer a high therapeutic potential, since together they are particularly important in the oxidation of lipids and other free radicals that are very active physiologically [32, 24].

Consumption of the ripe pseudo-fruit is recommended both for its flavour qualities and medicinal properties, as well as for its high nutritional value. It is important to note that its vitamin C content (203 mg/100g) is four times higher than that of guava and oranges, making it traditionally known as the main source of this nutrient, in addition to its minerals and other vitamins. One glass of cashew juice fulfils all of an adult's daily vitamin C needs [6].

Cashew cultivation has medical applications; its oil was proposed and applied for the first time in the country in 1868 in the treatment of leprosy skin lesions by Dr L. D. Beauperthuy. In addition, the juice of the cashew pseudofruit is digestive, bactericidal and anti-dysenteric, with tannins and minerals, making it a first-rate tonic, as well as detoxifying, anti-enteric and diuretic [6].

The bark of the tree is used for medicinal purposes, used in tea as a remedy against diarrhoea, swelling of the joints caused by syphilis and against diabetes. It is also rich in tannins, which makes it astringent and is used for tanning leather.The juice of the bark and the oil of the nut are considered remedies against corns and warts, cancerous ulcers and even elephantiasis. On the other hand, anacardic acid, present in the seed coat, has been shown to have some activity against "Walker 256 carcinoma". In addition, the oily substance of the follicle is used against cracks in the feet.In Cuba, they used the bark in herbal tea to treat asthma, colds and congestion. Chestnut oil is considered good against amoebas and is used to treat gingivitis, malaria and syphilitic ulcers. In addition, in Indonesia, the old leaves are used as a poultice on burns and skin diseases, and its juice is used to treat angina and, in the Philippines, dysentery.

1.7 Standardisation and quality inspection

Nowadays, the commercialisation of agricultural products, especially fruit and vegetables, requires products of the highest quality. Quality is the set of characteristics that qualify the food, within four general parameters: health, nutritional value, organoleptic or sensory characteristics and physical-mechanical properties, which are necessary for direct human consumption or for its processing and industrial transformation [33].

There are three factors that influence fruit quality. One is the varietal factor, which determines the presence of very marked physicochemical differences between plants, allowing us to know the commercial potential of each variety, whether for fresh consumption or for use on an industrial scale. Another factor is the external environment, where temperature, relative humidity, light intensity, soil texture, wind and rainfall, among others, play a significant role in fruit quality, depending on the type of fruit and the variety. Finally, the agronomic management of the plantations, which includes pruning practices, irrigation, planting density, pest, disease and weed control, can favour or affect the quality of the final product [34].The indices used internationally to assess the physicochemical and microbiological quality of fresh and processed fruit are varied, depending on the characteristics of the fruit and vegetables, the characteristics of the product to be obtained from them and their intended use. These indices make it possible to establish the degrees of quality in the different products, their use and value [33].

CHAPTER II
MATERIAL AND METHODS

2.1 Test site and characteristics of planting

The research was carried out on a cashew tree plantation (Anacardium occidentale L.), located at CORPOZULIA's Socialist Centre for Fruit Research and Development (CESID-Frutícola e Apícola), situated at km 27 on the road to "San Rafael de El Moján", in the north-western region of the Maracaibo plain (11°00'N, 71°30'O). This area has been classified as a tropical very dry forest life zone, with very scarce water resources, stretching from sea level to around 600 metres above sea level. Annual rainfall in the region ranges from 500 to 600 mm, with an irregular rainfall distribution regime that graphically corresponds to a bimodal curve. The average annual temperature ranges from 23 to 29°C, with a relative humidity of 75 per cent and an annual evapotranspiration of 2,000 to 2,200 mm [35, 36].

The two peaks of maximum precipitation occurred in May (45 mm) and November (212 mm), while the two peaks of minimum precipitation were in September (12 mm) and December (7.6 mm), with an annual average of 519.1 mm; an average temperature of 28.8°C and a relative humidity of 78.9%. The peaks of maximum rainfall were in June (145 mm) and October (57 mm); the peaks of minimum rainfall were in January (4 mm) and September (10.5 mm), with an annual average of 348.5 mm; the average temperature was 29.7°C and the average relative humidity was 76.4%. (Meteorological station of the Zulia State Viticultural Centre. CORPOZULIA).

The cashew pseudofruits were collected from a population of 20 plants, 10 of which belong to the yellow type and 10 to the red type, from a selection of 406 plants grown on an area of 2.5 ha. These plants are 2 years old, with agronomic management that includes fertilisation with a complete formula, cleaning pruning, weed, pest and disease control, and no irrigation programme. Five fruits per plant were harvested by hand in two different production periods: February-March and July-September, in the morning.Four collections were made per period, totalling 800 samples. Once the fruit had been harvested, they were placed in plastic bags marked with the mother plant number, the colour of the cashew pseudofruit, the harvest period and the sample number, to be transported in a cold room to the Post-Harvest Laboratory at CESID - Frutícola e Apícola.

2.2 Preparation of samples

Once in the laboratory, the fruit was washed with running water to remove dust and reduce the internal temperature, which favours ripening after harvest. They were then dried to begin the physical characterisation, which included determining weight with and without nuts, size and firmness.

2.3 Determination of quality parameters physical

The physical characterisation of mass, firmness and volume was carried out on whole fruit, with and without the nut.

2.3.1 Pasta

An electronic balance (Mettler PC4400) was used to measure the mass of each fruit, with and without the nut, and determine the average mass of the samples. The results were expressed in g/fruit.

2.3.2 Firmness

Firmness, measured as 1/10 mm of penetration, was determined by a puncture exerted on the epidermis or rind of the cashew pseudofruit using a penetrometer, model Universal Tester, brand Humboldt MFG Co™ (Norridge, Chicago, Ill, 60656),with a No. H-1240 needle, which has an inverse relationship with penetration. Thus, the greater the penetration, the less firm the fruit.

2.3.3 Volume

The cashew pseudofruit samples were pressed in a manual juice extractor to obtain the volume (mL), which was measured in a graduated glass cylinder with a capacity of 100 mL.

2.4 Determination of quality parameters chemical

The chemical characterisation of pH, titratable acidity, total soluble solids and ripeness index was carried out on the juice obtained from the pseudo-fruits of the cashew tree.

2.4.1 pH (acidity ionic)

It was determined according to the method described in standard 1315-79 of the Venezuelan Commission of Industrial Standards (COVENIN, 1979) [37], using a Metrohm™ model 744 potentiometer. The results were expressed in pH units.

2.4.2 Acidity titratable

The determination was carried out in accordance with COVENIN Standard 1151-77 (COVENIN, 1977) [38]. The results were expressed in grams of malic acid per 100 mL of sample.

2.4.3 Total soluble solids (°Brix)

The total soluble solids content, or ºBrix, was calculated in accordance with COVENIN Standard 1151-77 (COVENIN, 1977) [39] and determined using a manual refractometer or brixometer (Komax brand), with a scale of 0 - 32º.

2.4.4 Index from maturity

The maturity index was determined by the ratio between the total soluble solids content and the titratable acidity.

2.5 Production process for products derived from cashew pseudofruits

The products derived from the cashew pseudo-fruit in this scientific research were made in the post-harvest laboratories at CESID-Frutícola and Apícola.

2.5.1 Process for producing pasteurised juice from the pseudo-fruits of cashews

The raw material for the processing of cashew pseudofruit juice begins with the reception of the cashew fruits. The juice was made following the General Standard for Fruit Juices and Nectars. CXS 247-2005. Codex Alimentarius Commission, adopted in 2005 and amended in 2022 [42]. A scale was used to weigh the fruit, which was then washed with chlorine at 2 ppm for 3 minutes. The fruit was then sorted and the nuts separated from the fruit, leaving the pseudo-fruit to follow the cashew juice production line. The cloudy juice from the pseudo-fruits was obtained using an Oster® model FPSTJE316W juice extractor. After extraction, the cloudy juice was clarified using 1 g of commercial unflavoured gelatine at room temperature. The mixture was stirred and left to stand for 15 minutes until coagulation occurred on the surface of the liquid.Once clarified, the juice was filtered through a cotton gauze strainer, discarding the coagulated phase. The juice was pasteurised for 30 minutes at a temperature of 85 °C.

2.5.2 Process for producing nectars from cashew pseudofruits

The processing of cashew pseudofruit nectars, following the General Standard for Fruit Juices and Nectars. CXS 247-2005. Codex Alimentarius Commission, adopted in 2005 and amended in 2022 [42], is carried out in a tank with a mixture of: 1. stabiliser with sugar and dilution with hot water (60 °C-70 °C) and 2. Cashew pseudofruit pulp concentrate, sugar and acidulant boosters. Immediately, the pH, acidity and °Brix are adjusted, and de-aeration and pasteurisation are carried out at 80 °C - 90 °C for a period of 15 - 16 minutes. The pasteurised nectar is then filled as follows: 1. Cooled to room temperature (Tetra Brix) and stored at room temperature; 2. Hot filled (Bottle/Can) and stored at room temperature; and 3. Cooling to a temperature of 4 °C - 6 °C and storage (4 °C - 8 °C) in packaging (PURE PAK/Plastic).

2.5.3 Process for producing cashew pseudofruit liqueur

The process of making liqueur from cashew pseudofruit starts with the cloudy juice, as indicated in part 2.6.1.
Procedure for making cashew pseudofruit juice. The following steps are carried out: dilution, addition of enzymes, disinfection, clarification and fining of the must. Immediately afterwards, the must is left to rest for 24 hours at a temperature of 2°C.Once the wort has rested, it is filtered and the chaptalisation process is carried out, where the vat foot is prepared using Saccharomyces cerevisiae yeast. Immediately, a resting phase begins for 24 days at a temperature of 37 °C. After this phase, the process of filtering, sterilising, bottling and storing the liquor begins [6].

2.5.4 Process for producing jelly from the pseudofruits of cashews

The jelly is a product formulated from cashew pseudofruit and sugar, as suggested by [44, 45, 46]. The pH of the mixture should be adjusted by adding an acidifier such as citric acid. It is also necessary to increase the pectin content of the mixture by adding citric or malic pectin to obtain a suitable gel.Unlike jellies that are made with the whole fruit and all its pulp, in this research they were prepared with the cloudy juice of the cashew pseudofruit, adding sugar, pectin and citric acid.The jellies are formulated with cloudy cashew pseudofruit juice and refined white sugar in a 50-50 ratio. The gel is formed when the mixture reaches 65 °Brix (65% sugar), an acidity of 1% and a total pectin content of 1%.Cooking the mixture is the operation that has the greatest impact on the quality of the jam; it therefore requires a lot of skill and practice on the part of the operator. The cooking time depends on the variety and texture of the raw material. Therefore, a short cooking time is very important to preserve the natural colour and flavour of the fruit, while overcooking can cause the jam to darken due to the caramelisation of the sugar. The mixture was cooked for 30 minutes.The jam was filled into previously sterilised glass jars. The jars were washed with water and detergent, rinsed with plenty of water and boiled in water for 10 minutes. They were then removed and placed on a tray previously washed with hot water and covered to avoid contamination, avoiding hand contact with the rim and inside of the sterilised jars. The jam was filled at a temperature above 85 °C and closed with sterilised lids. The jars were cooled with water to room temperature and dried. They were left to stand for 24 hours. They were then stored in boxes in a clean, dry and cool place.

2.5.5 Process for producing jam from the pseudofruits of cashews

The stages of the jam-making process are described below:

• **Selection of** the **Raw Material**: Selection of cashew pseudo-fruits that are in good condition, discarding those with defects.

• **Washing**: The process of making jam begins with washing the cashew fruits. They should be washed with water and a chlorine solution at a concentration of 2 ppm. The selected fruit must be free of mechanical damage caused by animals (such as bird beaks, mishandling, etc.) and the visible presence of microorganisms (such as fungi and yeasts). The fruit must be left in the disinfectant solution for three minutes to ensure the elimination of microorganisms and/or contaminants on the skin of the pseudofruit. After this time, the fruit is rinsed with plenty of water to remove any residues of the disinfectant solution.

•**Separation**: This stage involves separating the nut from the pseudo-fruit with gentle rotating movements using the hands. The separated nuts are placed in a clean container to be stored, as they can be used to make other products. The pseudofruits are weighed on a scale and the weight is recorded.

•**Pulping**: This involves obtaining the pulp from the cashew pseudofruit. To do this, the whole pseudofruit is processed using an Oster® model FPSTJE316W juice extractor. This results in fibre-free juice on the one hand and fibres and residues as surplus on the other. These surpluses can be used then used to make other by-products, such as flours, preserves, etc.

•**Weighing the ingredients**: All the ingredients in the jam formulation, such as the pseudofruit pulp, sugar, carboxymethylcellulose (CMC) thickener and lemon juice, are weighed on the scales according to the proportions in the formulation.

•**Sweetening**: Add sugar to the puree in the appropriate proportion, according to the specific recipe.

•**Cooking**: All the ingredients are placed in the stainless steel tank and the stirring system is activated, which must be constant throughout the cooking process to prevent the thickener from clumping. The cooking temperature is set to operate within the range of 80 to 100 °C, and the mixture is cooked with constant stirring until it begins to boil. Cooking continues until the mixture reaches the thickness and consistency characteristic of a jam, in accordance with the standards established by the Venezuelan COVENIN (COVENIN 2005:1994) [46] and Codex Alimentarius (CODEX STAND 79-1981) [47] standards. Before finalising, add the lemon juice and mix well.

•**Acidity adjustment**: Addition of citric acid or another acidifier to adjust the pH and improve the preservation and flavour of the jam.

•**Packaging**: While still hot, the jam is placed in sterilised glass jars, which are immediately sealed with lids.

•**Sterilisation:** The jam jars that have already been filled are taken to an autoclave for sterilisation. After the sterilisation time, the jars are removed from the autoclave and left to cool at room temperature. Once cool, the jars with jam should be stored at room temperature in a cool, clean place, as cashew pseudofruit is very sensitive to browning caused by the formation of 5-HMF (hydroxymethylfurfural) at high temperatures and long heating times.

•**Cooling and labelling**: After sterilising the bottles, they must be left to cool at room temperature before being labelled, because if they are not, the labels will not adhere. The information on the label must contain everything required by the

Venezuelan standard COVENIN 2952:2001 (1st Revision) [48], which establishes the general standard for labelling packaged foods.

2.6 Design and statistical analysis of results

For the statistical analysis, a completely randomised experimental design was used with a 22 x 4 x 10 factorial arrangement, in plots divided in time, with a hierarchical sampling effect within the harvest period. An analysis of variance was carried out using the PROC-GLM method and a separation of means using the LS MEANS procedure of the Statistical Analysis System (SAS, 2000) [49].

CHAPTER III
RESULTS AND DISCUSSIONS

3.1 Features physical

The results of each variable of the physical characteristics are discussed below by the effects analysed.

3.1.1 Effect of the type of cashew

The statistical analysis of the results of the physical characteristics in this research showed significant differences ($P<0.05$) only for the mass of the pseudofruit without a nut, fresh mass yield and firmness between the yellow and red types, not for the mass of the pseudofruit with a nut between the yellow and red types (see Table 3.1).

Table 3.1

Variation in physical characteristics between yellow and red cashew (Anacadium occidentale L.) pseudofruits with and without nuts

	Type		
Variable	Yellow	Red	EP
Pasta with walnuts (g)	61.83a	57.56a	1.43
Nut-free pasta (g)	56.25a	52.08b	1.28
Fresh mass yield (%)	90.66a	90.05b	0.15
Firmness, 1/10 mm	93.73a	80.80b	2.39
a,b Different letters on the same line indicate differences ($P<0.05$). EP: Error standard			

In general, the mass ratio between the stalk and the nut is 9:1, with 90 per cent representing the stalk and 10 per cent the nut [50]. Meanwhile, the ratio of nut to kernel (edible part) ranges from 20.1% to 27.2% [51]. Pioneering reports on research in the eastern region of Venezuela determined a lower average mass for the pseudofruits of the red cashew tree (47.5 g) and a very similar one for the pseudofruits of the yellow cashew tree (60.1 g) [6]. Avilán et al. (1989) reported values similar to those of the present study, establishing that the average mass of the pseudofruit varies between 44 g and 137 g, corresponding to a nut of 5 g to 7 g; however, they did not discriminate between the types of cashew tree analysed [52].In Colombia, it has been reported that the average mass of a walnut is 6.5 g, while the pseudofruit weighs an average of 80 g [1], which is higher than the value obtained in this study. However, the results of Bezerra et al. (2018) in Brazil indicated a range of pseudofruit mass between 45 g and 108 g, a range in which the materials evaluated in this study are found [53].

Table 3.1 shows that firmness is higher in red cashew trees ($p<0.05$) compared to yellow cashew trees. Firmness is directly related to fruit quality. From an agro-industrial point of view, this aspect gives the red type greater comparative advantages over the yellow type, since firmness determines better resistance to damage caused by transport and handling, as well as less susceptibility to attack by micro-organisms. This fruit characteristic depends on the effect of enzymes on pectin and starch during the ripening process. Protopectin is degraded into lower molecular weight and more water-soluble fractions, which causes the fruit to soften [50]. As ripeness increases, cell rigidity softens, increasing sensitivity to mechanical damage and microorganisms.However, cashew nuts generally have a delicate constitution that makes them extremely perishable due to the lack of a consistent coating and their high water content, which is around 86 per cent [54].

With regard to the chemical characteristics of the juice from the types of pseudofruit assessed, the analysis of variance only detected differences in the ripeness index (Brix/titratable acidity), with the red type showing the lowest values (see Table 3.2).As for density, the values obtained are slightly higher than those tabulated by Alejos [6], which varied between 0.95 and 0.96.When analysing the titratable acidity variable, in terms of the percentage content of malic acid, similarities were observed with the values reported in the literature, which vary between 0.36 and 0.44 mg/100 ml of juice, and which explain the astringency that characterises them, due in part to the content of a large proportion of organic acids and tannins. These low pH values are a consequence of a high concentration of organic acids (citric, ascorbic and malic acid), which are commonly used in the food industry to preserve flavour and prevent the attack of microorganisms, especially pathogenic bacteria [6].

Table 3.2

Variation in the chemical characteristics of the juice of yellow and red cashew pseudofruits (Anacadium occidentale L.)

		Type	
Variable	Yellow	Red	EP
Juice yield (%)	53.38a	54.47a	0.80
Density (g/mL)	1.05a	1.06a	0.002
pH	4.20a	4.27a	0.04
°Brix (%)	13.56a	13.48a	0.09
Titratable acidity (%)	0.33a	0.32a	0.01
Maturity index	52.41a	47.23b	1.12
a,b Different letters on the same line indicate differences (P<0.05). EP: Error standard			

The total soluble solids content (°Brix) was slightly lower than that obtained in yellow (14.11 %) and red (14.24 %) cashews from older Venezuelan plantations, but higher than the average value reported in Brazilian studies (10 % - 12 %) [6].The Brix/titratable acidity ratio was significantly ($p<0.05$) higher in the yellow pseudofruits. The values found for both types duplicate the maturity indices reported in pseudo-fruits from older plantations grown in the eastern region of the country, which showed the opposite behaviour to those reported in this research, when the red type showed the highest indices. They are also higher than the average values (30.6) found in cashew pseudofruits grown in Brazil [6].

3.1.2 Effect of selected cashew trees

Significant differences ($p<0.05$) were only observed in the total soluble solids content (°Brix) between the plants (see Table 5). The results are shown in Figure 1.The total soluble solids content is a chemical parameter used to determine the degree of maturity of the fruit, along with other physical parameters (ease of abscission) and some visual means (skin colour). These last two parameters were used in this research as criteria for picking the fruit.It is also possible that the state of maturity of the fruit at the time of harvest was not the same, resulting in the aforementioned differences. However, the variations due to a non-uniform state of maturity when the fruit was harvested were not reflected in the maturity index (Brix/titratable acidity), which was not affected by the plant factor. Given that the agronomic management of these plants was similar, as was their age, it is possible that the variability observed in the total soluble solids content is due to their genetic variability as they were obtained by seed propagation. Seed propagation of Anacardium occidentale L. is said to be the easiest, but it is not the most recommended, as the plants obtained produce unequal fruit, both in type and in quality and quantity [6].

Table 3.3

Descriptive statistics of some physico-chemical parameters of cashew pseudofruits (Anacardium occidentale L.) according to the trees selected in this study.

Variable	Average	Maximum	Minimum	Deviation standard
Pasta with walnuts (g)	59.89	83.47	40.12	11.58
Nut-free pasta (g)	54.17	75.91	35.04	11.12
Fresh dough yield (%)	90.36	93.45	85.82	1.99
Firmness, 1/10 mm	87.26	115.04	46.52	19.39
Juice yield (%)	53.92	66.11	42.69	6.65
Density (g/mL)	1.05	1.08	1.02	0.015
pH	4.23	4.81	3.57	0.32
°Brix (%)	13.52	16.24	11.02	1.49
Titratable acidity	0.32	0.6	0.17	0.13
Maturity index	49.82	86.43	25.04	

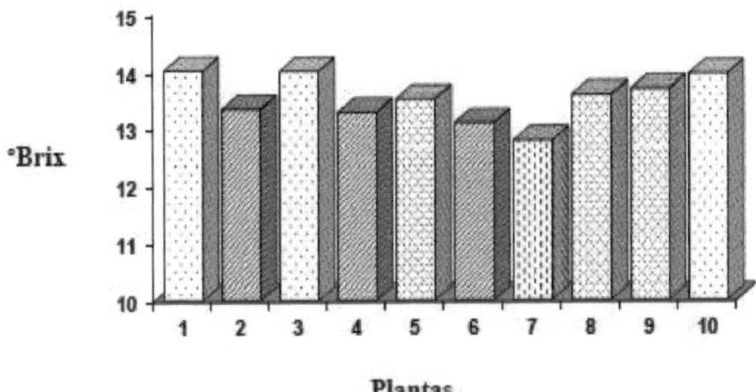

Figure 3.1 Variation in the total soluble solids content of cashew (Anacardium occidentale L.) pseudofruits as a function of plant selection

The different changes that occur during the organoleptic ripening process (colour, odour and consistency) appear to be synchronised and controlled by a genetic factor, based on the fact that the interval between anthesis (flower opening) and organoleptic ripeness under similar climatic conditions is relatively constant for a given fruit [6].

3.1.3 Effect of harvest time

When studying the effect of the harvesting period on the physicochemical parameters of cashew pseudofruits, it was observed that only the density of the juice did not vary ($p>0.05$).Table 6 shows that the highest values for all the variables were obtained in the harvest period between February and March, which corresponds to the dry season, i.e. when there was no rainfall. Several authors agree that the fruit set of the cashew plant is highest when the sunlight levels are highest. Theoretically, sunlight directly influences fruit quality due to its important role in carbohydrate synthesis [6]. In addition, high temperatures produce variations in the total soluble solids content. Thus, it has been shown in non-climacteric fruit trees that temperature plays a very important role in the ratio between total soluble solids and organic acids. It was concluded that the recommended average temperature is 26 °C [6].

Table 3.4

Variation in the physicochemical characteristics of cashew pseudofruits (Anacardium occidentale L.) according to harvest time

Period			
Variable	FEB - MAR	JUL - SEPT	EP
Pasta with walnuts (g)	61.70a	57.69b	1.178
Nut-free pasta (g)	56.52a	51.81b	1.132
Fresh mass yield (%)	91.19a	89.52b	0.189
Firmness, 1/10mm	94.41a	80.12b	1.517
Juice yield, %	56.38a	51.46b	0.595
Density, g/mL	1.058a	1.049a	0.003
pH	4.08a	4.38b	0.028
°Brix (%)	14.16a	12.88b	0.143
Titratable acidity (%)	0.405a	0.236b	0.009
Maturity index	39.41a	60.24b	1.304
a,b Different letters on the same line indicate differences (P<0.05) EP: Standard error			

3.1.4Effect of sampling on harvest time

All the variables studied, with the exception of density, were affected ($p<0.05$) by fruit sampling within each period considered in this study. The results are shown in figures 2 to 10.With regard to the mass of pseudofruits with and without nuts, the same trends or behaviour can be observed (Figure 2 and Figure 3). During the first season (February - March), the pseudofruits experienced a slight increase in mass, with no appreciable significant differences ($p>0.05$) in each sampling carried out; in contrast, during the second season (July - September), their mass tended to decrease throughout the samplings, with marked differences ($p<0.05$) in samplings II and IV. It was mentioned earlier that atmospheric conditions, including CO2 concentration, relative humidity and temperature, are closely related to yield. In this sense, it can be seen that the pseudofruits with the highest mass were grown when the weather conditions in the Mara region (Venezuela) corresponded to an average temperature of 28.2°C, a wind speed of 4.8 km/h in the absence of precipitation, contrasting with the weather conditions observed during the second harvest season, which were higher. However, there were no marked differences in relative humidity between the two periods (74.5% and 74.7%). The results are in line with reported studies, which conclude that the recommended average temperature is 26°C, with a season marked as dry, followed by a rainy season.

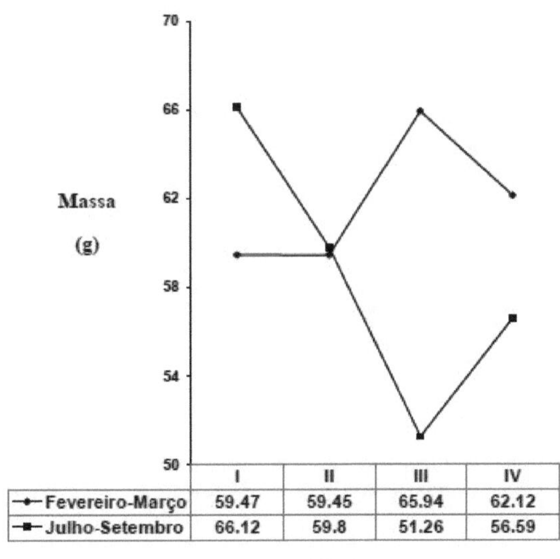

Figure 3.2 Variation in the mass of walnut pseudofruits between sampling periods

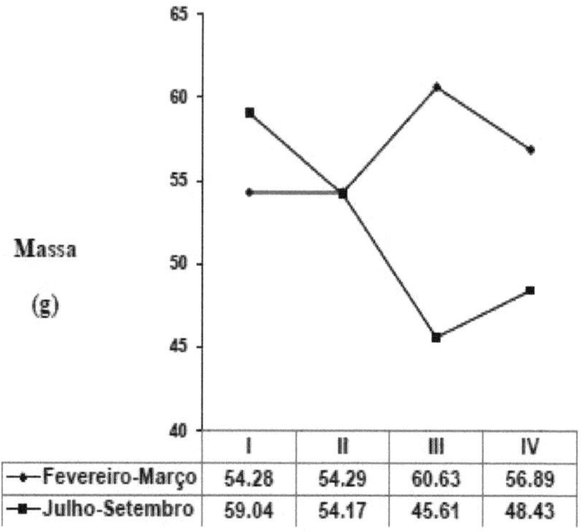

Figure 3.3 Variation in the mass of pseudofruits without nuts between sampling periods

The events observed above are reflected in the % yield in the mass of the pseudofruits without nuts (Figure 4). Similarly, the best yields were always obtained during the period from February to March, maintaining an average % mass of over 90%.

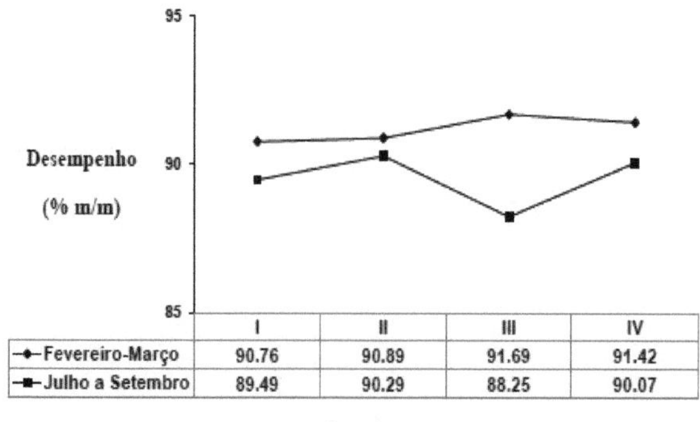

Figure 3.4 Variation in pseudofruit fresh mass yield between sampling periods

The firmness of the pseudofruits also varied ($p<0.05$) between the samples in each harvest season (Figure 5). During the first season, the pseudofruits maintained an average firmness of 94.4/10 mm, and there were no significant changes ($p<0.05$) between samplings. In the period from July to September, two groups of different averages were observed; in the first samplings, fruit were harvested that showed an average firmness of 89.6/10 mm, while in the final samplings the average was 70.7/10 mm, resulting in firmer fruit. The lower firmness of the fruit harvested in the different samplings of the first season seems to be influenced by the accumulation of water in the soil due to the rains that occurred in the period (386.1 mm of precipitation) before harvest, which caused an increase in turgidity and greater susceptibility to mechanical damage, reducing their quality [6].

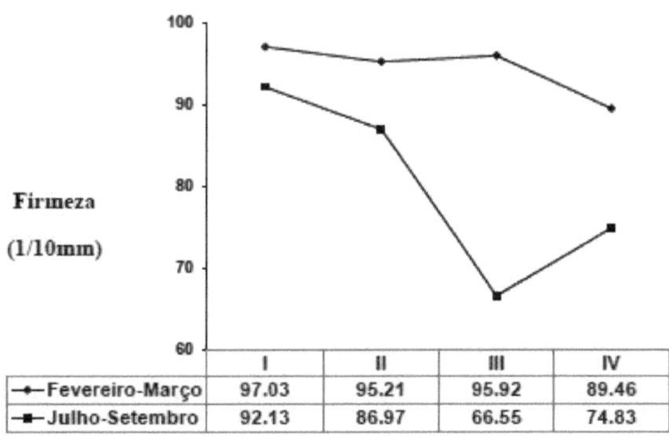

	I	II	III	IV
Fevereiro-Março	97.03	95.21	95.92	89.46
Julho-Setembro	92.13	86.97	66.55	74.83

Figure 3.5 Variation in pseudofruit firmness between samples at each time period

The above about firmness and the accumulation of water in the soil before harvest is partly corroborated when analysing the juice yield (Figure 6).

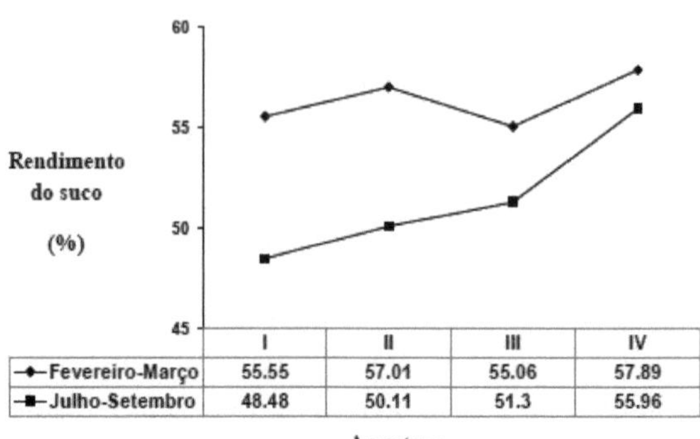

Figure 3.6 Variation in pseudofruit juice yield between sampling periods

The water content of fruit and vegetable products is closely related to the humidity content of the surrounding environment. It has been observed in non-climacteric fruit that an increase in the relative humidity of the environment and sufficient irrigation or a high accumulation of water in the soil simultaneously produce a decrease in firmness and an increase in juice yield. These theoretical foundations coincide with the results obtained in this research [6].

When the changes in ionic acidity (pH) were analysed (Figure 7), it was noted that there were no significant changes ($p>0.05$) during the samples taken between February and March, with the average value concentrated at 4.08. This value was approximately 6.7% lower than the average value for the period from July to September, when there were differences ($p<0.05$) in the pH of the pseudofruits between the samplings intermediate. During the second harvest, the pseudofruits were slightly less acidic.

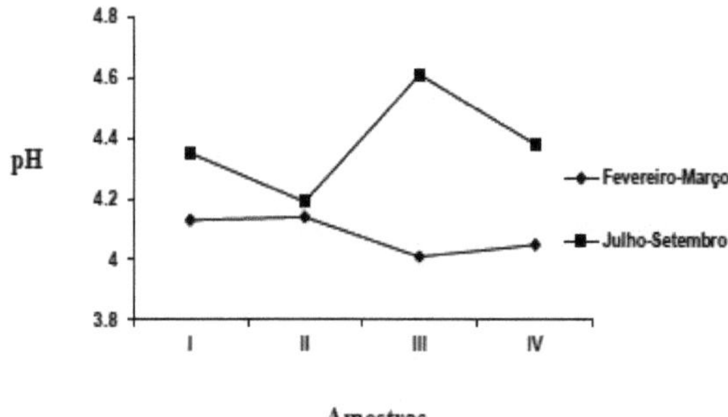

Figure 3.7 Variation in the pH of the pseudofruits between the samples in each time period

The period from February to March was characterised by a total lack of precipitation, unlike the period from July to September. It is possible that the humidity in the environment during the second sampling period had a diluting effect and greater mobilisation of the chemical components of the pseudofruits, causing a decrease in the content of organic acids and sugars and a lower °Brix reading (Figure 8).

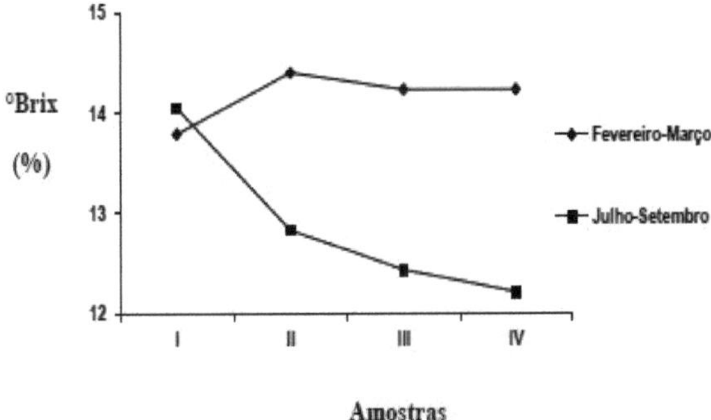

Figure 3.8 Variation in °Brix in the pseudofruit juice between samplings in each time period

Field experiments in Brazil have shown that in the rainy season there is a risk that cashews will be less flavoursome due to the high moisture content, which has a diluting effect on soluble solids. However, it was found that the pH values of the juice of the pseudofruits fluctuate between 3.5 and 4.6, regardless of the time of harvest, agreeing with the values obtained in this research study [55].
The aforementioned dilution effect produced by changes in the humidity of the environment was also reflected in the titratable acidity values, expressed in terms of the malic acid content in the juice (Figure 9).

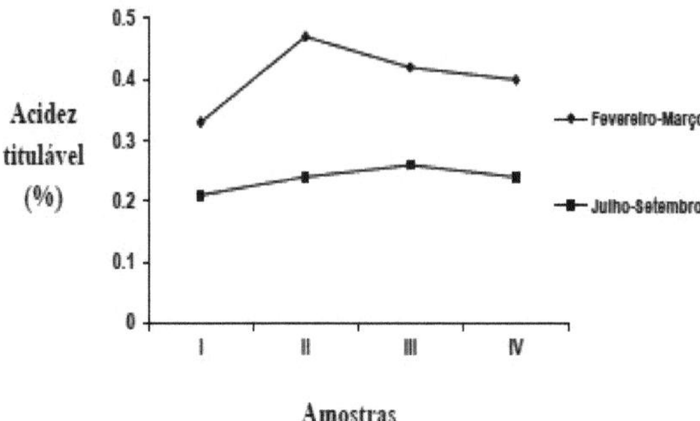

Figure 3.9 Variation in titratable acidity of pseudofruits between sampling periods

There were no differences (p>0.05) during the July to September samplings, with an average value of 0.24 per cent, significantly lower than the average value observed during the February to March samplings (0.41 per cent), where there were significant changes (p<0.05) between the samplings.
Given that there is an inverse relationship between the maturity index and the titratable acidity of any fruit, it is to be expected that, with a higher proportion of malic acid in the fruit harvested from February to March, the maturity indices would be lower (Figure 10). This aspect seems to be contradictory, since these fruits contained a higher proportion of total soluble solids [55].

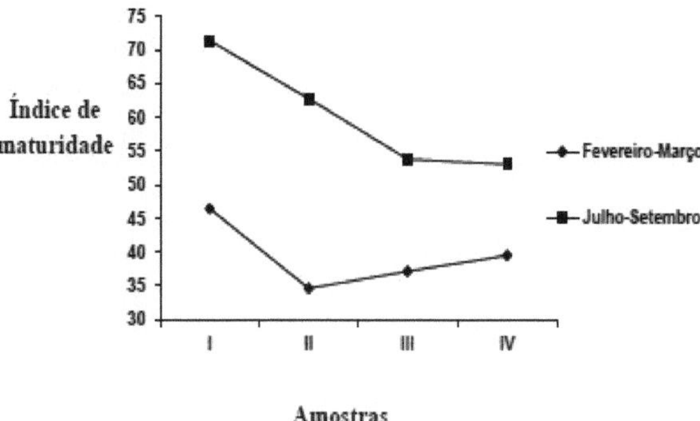

Figure 3.10 Variation in the maturity index of the pseudofruits between the samples in each time period

Although all the independent effects on each of the variables considered in this research have been discussed, the analysis of variance showed that some of them were also affected by the interaction between some independent effects, which undoubtedly conditions and modifies their behaviour, as will be shown below [55].

3.1.5 Interaction effect Type * Time

The analysis of variance detected differences ($p<0.05$) only for the firmness, titratable acidity, juice yield and maturity index variables. As mentioned, firmness and resistance to penetration have an inverse relationship. As such, Figure 11 shows that the pseudo-fruits of both types of cashew were always less firm ($p<0.05$) during the harvest period from February to March, with no differences between the types. During the second period (July to September), it was observed that the yellow-coloured fruits were always the least firm, i.e. those with the least resistance to penetration, compared to the red pseudo-fruits.

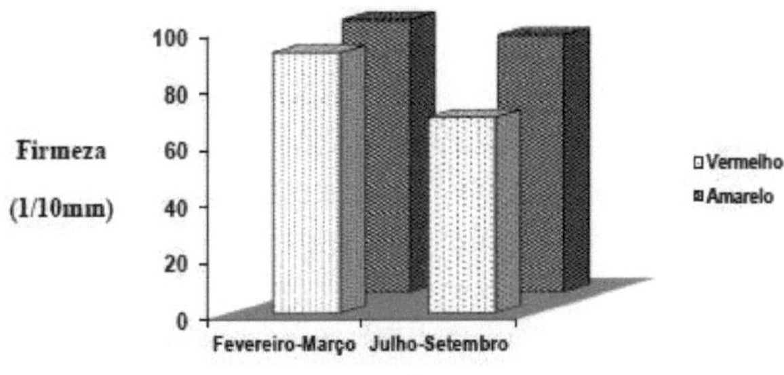

Figure 3.11 Variations in the firmness of yellow and red cashew pseudofruits depending on harvest time

The loss of firmness is a consequence of the loss of turgidity of the cells and is related to the state of maturity and the water content present in the fruit, causing the fruit to become flabby and soft [6, 55]. The susceptibility of pseudo-fruits harvested from February to March to water accumulation in the tissues is corroborated by the values obtained for juice yield (Figure 12). Thus, those with the lowest juice yield were the firmest pseudofruits, i.e. those harvested in the second harvest period (July to September).

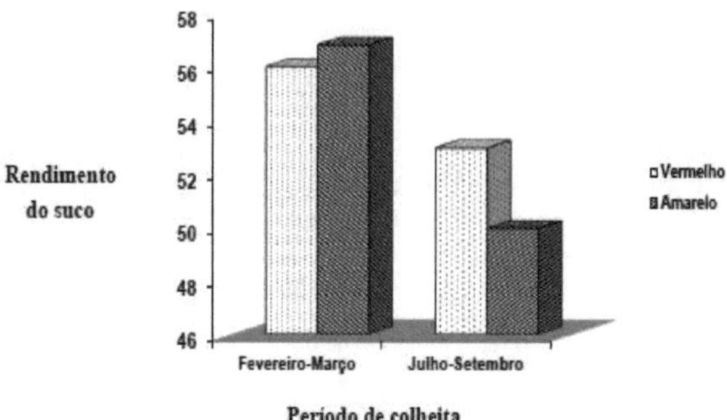

Figure 3.12 Variations in the juice yield of yellow and red cashew pseudofruits, depending on the time of harvest.

The environmental factors considered to explain the behaviour of these same variables due to the sampling effect at each harvest time justify the differences observed here. As for the titratable acidity of the juice (Figure 13), it was observed that the pseudofruits harvested between July and September had lower concentrations of malic acid, which is in line with their higher maturity indices when compared to the values obtained from the pseudofruits harvested between February and March (Figure 14). At the same time, there were drastic changes in the titratable acidity and maturity index of the yellow pseudo-fruits compared to the red ones. Thus, the titratable acidity of the yellow pseudofruits was 2.5 times lower in the period from July to September, corresponding to a ripeness index 2 times higher in the same period, compared to the period from February to March.

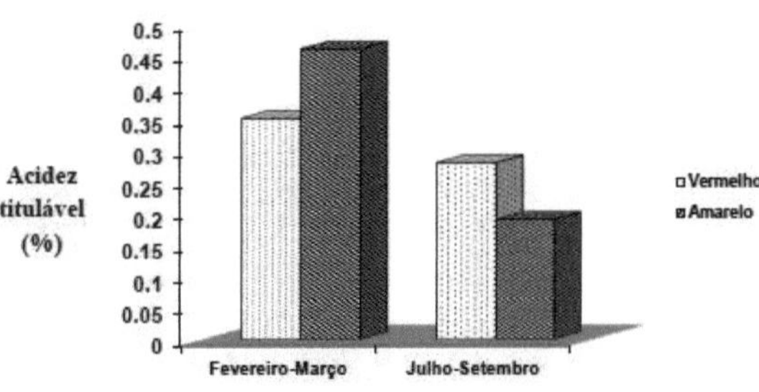

Figure 3.13 Variations in titratable acidity in yellow and red cashew pseudofruits, depending on the time of harvest.

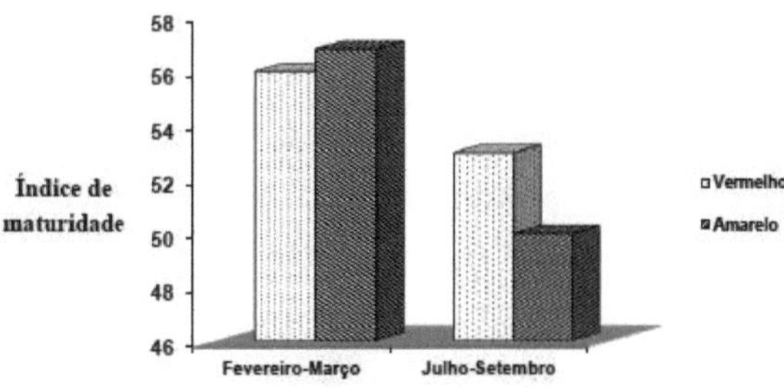

Figure 3.14 Variations in the maturity index of yellow and red cashew pseudofruits depending on the time of harvest.

3.1.6 Interaction effect Type * Sampling within harvest time

Significant differences ($p<0.05$) were detected in pseudofruit mass performance, firmness, juice yield, pH, total soluble solids content and titratable acidity, when evaluating the effect of the Type* Sampling interaction within each harvesting season. Figure 15 shows the changes that occurred in the mass performance of the pseudofruits in the different types of merey over the course of the samplings in the different harvest periods. During the first samplings, in both types of merey, no differences were observed ($p>0.05$), showing similar average yields (56.8 per cent and 55.97 per cent, for yellow and red, respectively). In contrast, changes There was a tendency for yields of both types to decrease, with the yellow-coloured types showing the lowest yields (52.9% and 49.9% for red and yellow, respectively).

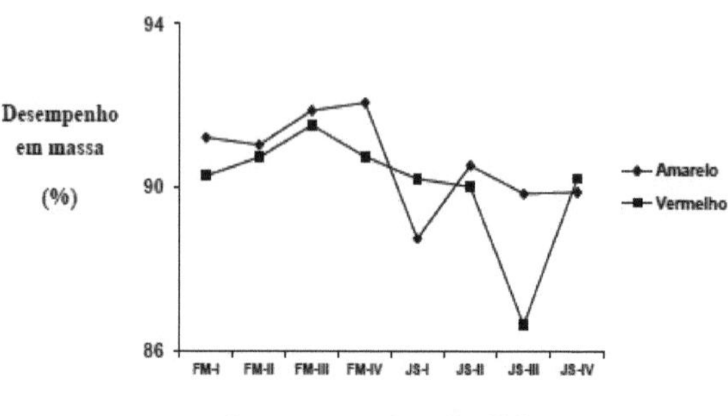

Figure 3.15 Variation in the percentage yield of yellow and red cashew pseudofruits in the different samples during the harvesting seasons.

A similar behaviour was observed when analysing the juice yield (Figure 16). Like the mass yield, the average juice yield of both types always tended to decrease in the samples from the second harvest period, when the most marked changes occurred. However, the average yields between the two types of merey were quite similar within the same period.

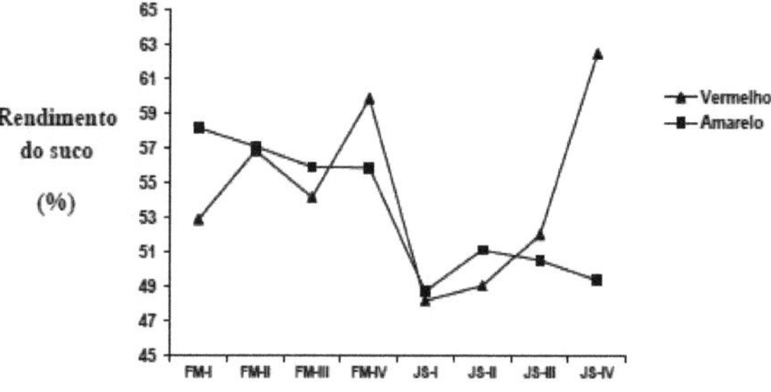

Figure 3.16 Variations in the juice yield of yellow and red cashew pseudofruits in the different samples during the harvesting seasons.

There were no differences (p>0.05) in the firmness of the pseudofruits between the two types in the different samplings carried out during the first period (February-March), maintaining an average of 97.22/10 mm for the yellow ones and 92.35/10 mm for the red ones. During the second harvest period, the differences became noticeable in the last samplings for both types of pseudofruit, with the red ones showing the highest resistance to penetration (Figure 17).

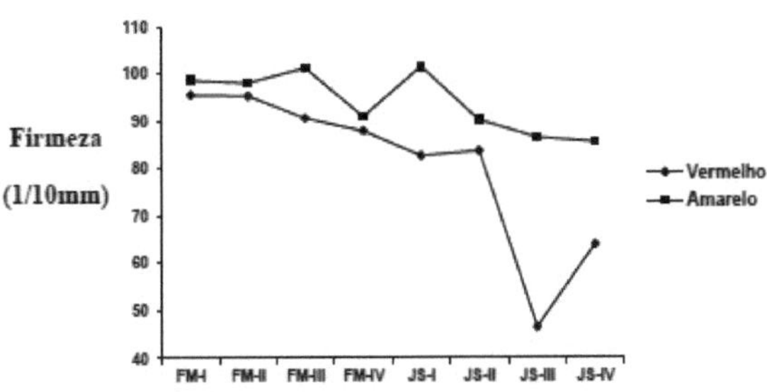

Figure 3.17 Variations in the firmness of yellow and red cashew pseudofruits in the different samples during the harvesting seasons.

The same trend was observed in the concentration of total soluble solids (Figure 18), with the greatest differences ($p<0.05$) between the yellow and red types in the last two sampling periods of the second harvest period. The interaction between the type and sampling effects in each of the harvesting periods considered determined differences ($p<0.05$) in the pH readings of the juice extracted from the harvested merey pseudofruits, with tendencies towards more basic values.

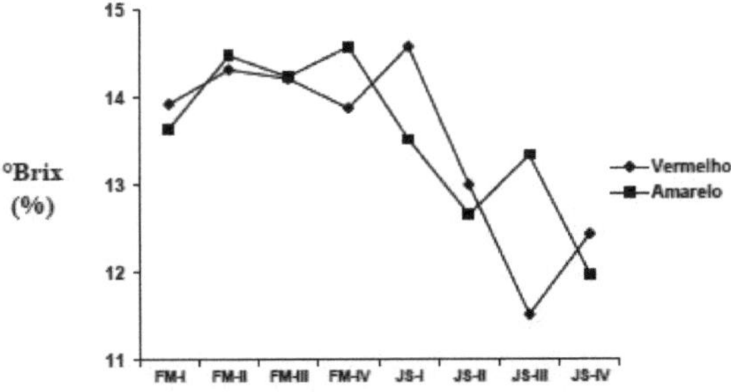

Figure 3.18 Variation in °Brix in yellow and red cashew pseudofruits, in the different samples during the harvesting seasons.

It can be seen in Figure 19 that the pH of the juice from the pseudo-fruits of both types of merey varied throughout the samples taken during the two harvest periods, with the yellow ones registering average values of 4.05 and 4.35, lower than the red ones, whose average values were 4.12 and 4.41 for the February-

March and July-September periods, respectively. The increase in pH values corresponded to a decrease in titratable acidity, due to the concentration of malic acid, as seen in Figure 20. The most notable changes ($p<0.05$) occurred in the yellow pseudo-fruits, where the average titratable acidity was 2.3 times lower in the second harvest period.

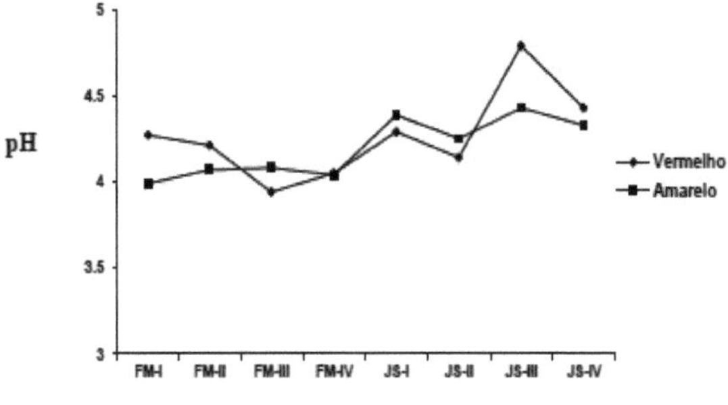

Figure 3.19 Variation in pH in yellow and red cashew pseudofruits, in the different samples during the harvesting seasons.

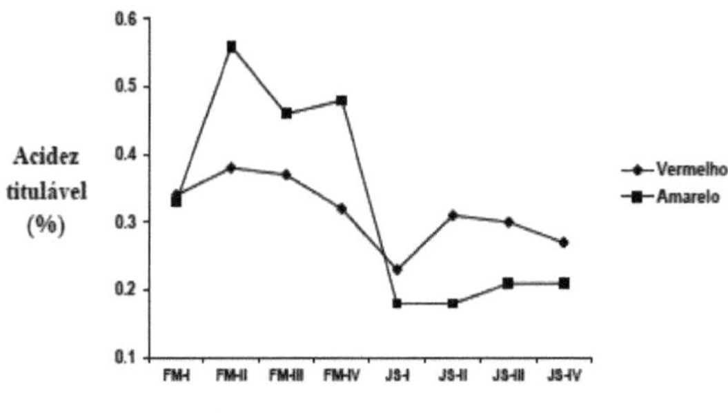

Figure 3.20 Variation in titratable acidity in yellow and red cashew pseudofruits, in the different samples during the harvesting seasons.

3.2 Evaluating the physical and chemical quality of cashew products

Different products were made from the juice of the merey pseudofruits of both types, including: pasteurised juice, nectar, merey liqueur, jam and compote (see Table 7).

Table 3.5

Physico-chemical characteristics of cashew (Anacardium occidentale L.) pseudofruit products

Quality parameter	Products					
	Pasteurised juice and nectar		Cashew pseudofruit liqueur		Jelly and jam	
	Yellow	Red	Yellow	Red	Yellow	Red
pH	4.20	4.27	3.21	3.46	3.8	4.2
°Brix	13.56	13.48	14.0	14.0	48.0	48.0
Titratable acidity	0.33	0.32	4.83	3.90		
SO2 Free	-	-	25.0 ppm	28.8 ppm	-	-
SO2	-	-	38.4 ppm	51.2 ppm	-	-
Total Alcohol content	-	-	8.4°	6.8°	-	-

These products are the most commercialised in Brazil, which is considered one of the countries in the American hemisphere that competes on the international market, along with India, in the industrialisation of this fruit. The results of the evaluations of the main quality parameters established and adopted internationally for these products are shown in Table 7. The values obtained for the products produced at laboratory level have physico-chemical characteristics similar to those of jams, jellies, juices, nectars and alcoholic beverages fermented commercially in Brazil and India. The jams and jellies made were subjected to tasting tests with untrained tasters using a hedonic scale. The results showed that the jams and jellies from the two types of merey were acceptable, with odour, taste, colour and appearance being classified in the "acceptable" to "good" range for the jams and jellies from yellow pseudofruits and "good" to "very good" for those from red pseudofruits.

CONCLUSIONS AND SUGGESTIONS FOR FUTURE WORK

The differences found between the physico-chemical parameters in this study are due to the variability in the chemical composition of the merey, which is mainly affected by factors such as: soil, irrigation, rainfall, type and ripeness of the fruit.The physicochemical characteristics are conditioned by the harvesting season in which they are obtained. Thus, pseudo-fruits harvested in the February-March period guarantee excellent quality raw material for industrial processing and the production of juices, jams and liqueurs. The high yield by weight of the merey pseudofruit, regardless of type, guarantees a high supply of raw material for industrialisation in the production of dried merey and merey in syrup.The physicochemical characteristics of the juice of the merey pseudofruit guarantee a high-quality raw material for the manufacture of bottled and pasteurised products (pasteurised juices, nectar, jam and compote) and its use in the production of liqueurs. However, their low levels of ionic acidity (pH) in the presence of a significant amount of sugars are favourable conditions for the development of contaminating microorganisms, which reduce the shelf life of the juice.The susceptibility of pseudofruits to mechanical damage, due to their low firmness, is one of the disadvantages of this fruit when it comes to post-harvest handling and transport. Carry out studies to mark or identify the fruit from anthesis or filling in order to determine the optimum harvest point that guarantees fruit with similar ripeness indices. Initiate studies on the conservation and storage of pseudofruits to guarantee the availability of raw materials for the industry.

BIBLIOGRAPHICAL REFERENCES

[1] Orduz, J. & Rodríguez, E. (2022). Cashew (Anacardium occidentale L.) a crop with productive potential: technological development and prospects in Colombia. Agronomía Mesoamericana, 33(2), 2215-3608. https://doi.org/10.15517/am.v33i2.47268

[2] de Abreu F. P., Dornier M., Dionisio A. P., Carail M., Caris-Veyrat C., Dhuique- Mayer C. (2013). Cashew apple (Anacardium occidentale L.) extract from by-product of juice processing: a focus on carotenoids. Food Chemistry, 138(1), 25-31. https://doi.org/10.1016/j.foodchem.2012.10.028

[3] Oliveira, N. N., Mothé, C. G., Mothé, M. G., & de Oliveira, L.G. (2019). Cashew nut and cashew apple: a scientific and technological monitoring worldwide review. Journal of Food Science and Technology, 57(1), 12-21. https://doi.org/10.1007/s13197-019-04051-7

[4] Prasertsri,P.,Roengrit,T.,Kanpetta,Y.,Tong-un,T.,Muchimapura,S., Wattanathorn, J., & Leelayuwat, N. (2013). Cashew apple juice supplementation enhanced fat utilisation during high-intensity exercise in trained and untrained men. Journal of the International Society of Sports Nutrition, 10(1), 1-6. https://doi.org/10.1186/1550-2783-10-13

[5] Cevallos, M., Urdaneta, F. and Jaimes, E. (2019). Development of agroecological production systems: Dimensions and indicators for their study. Revista de Ciencias Sociales (Universidad del Zulia, Venezuela), XXV(3), 172-185.

[6] Alejos-Pineda, R. E. (2003). Physico-chemical characterisation of the fruit of two types of merey (Anacardium occidentale L.) for industrialisation purposes.

[7] Alejos-Pineda, R., Arenas de Moreno, L., Ferrer, J., Castellano, G., Nuñez-Castellano,K. y Pérez-Pérez, E. (2022). Physico-chemical characterisation of the fruit of two types of merey (Anacardium occidentale L.) from a plantation in Mara, Zulia state, Venezuela. Revista Iberoamericana de Tecnología Postcosecha, 23 (2), 166-180.Available at: https://www.redalyc.org/articulo.oa?id=81373798007

[8] Sindoni, M., Loggiodice, P. R. H., & Natera, J. R. M. (2009). The merey (Anacardium occidentale L.): The fruity species of the Eastern Savas of Venezuela. Scientific Journal UDO Agrícola, 9(1), 1-8. Available en: https://dialnet.unirioja.es/servlet/articulo?codigo=3293540

[9] McLaughlin, J., Balerdi, C. y Crane, J. (2022). The marañón (Anacardium occidentale L.) in Florida. Horticultural Sciences, Florida Cooperative Extension

Service, Institute of Food and Agricultural Sciences, University of Florida (UF/IFAS),HS1041, 1-4. https://doi.org/10.32473/edis-hs291-2005

[10] Ramos, F., Osorio, C., Duque, C., Cordero, Aristizábal, F., Garzón, C. & Y. Fujimoto (2004). Chemical study of the giant marañón nut (Anacardium giganteum). Revista Académica de Colombia y Ciencia, 28 (109), 565-575.

[11] Sindoni, M, Marcano, L., & Parra, R. (2008). Acceptance studies of merey-derived harins for the preparation of breads. Agronomía Tropical, 58(1), 11-16. Available at: http://ve.scielo.org/scielo.php?pid=S0002-192X2008000100003&script=sci_abstract

[12] Pérez Morales, M.L. y Velázquez. F. (2021). Artisanal production of citric acid and ascorbic acid from caujil (Anacardium occidentale L.) and mango (Magnifera indica). (Special Degree Work for Chemical Engineers). Rafael Urdaneta University. Venezuela.

[13] Sindoni, V. M. J., Caldera, R. E., Pérez, A. C., Marcano, L., Parra, R., & Marín, R.C. (2007). Evaluation of coagulating agents for the formulation of juice from merey pseudofruits. Agronomía Tropical, 57, 61-65. Available at: http://ve.scielo.org/scielo.php?script=sci_arttext&pid=S0002-192X2007000100008

[14] Tuler, A. C., Peixoto, A. L. & Barboza da Silva, N. C. (2019). Unconventional food plants in the rural (UFP) community of São José da Figueira, Durandé, Minas Gerais, Brazil. Unconventional food plants in the rural (UFP) community of São José da Figueira,Durandé, Minas Gerais, Brazil. Rodriguésia, 70, e01142018, 1-12. https://doi.org/10.1590/2175-7860201970077

[15] Brito, E. S., Silva, E. O. & Rodrigues, S. (2018). Cashew - Anacardium occidentale. In S. Rodrigues, E. O. Silva & E. S. Brito (Eds.). Exotic fruits: reference guide. Academic Press.

[16] Sindoni, M., Hidalgo, P., Chauran, O., Chirinos, J., Bertorelli, M., & Salcedo, F. (2005). Merey cultivation in eastern Venezuela. Series manuales de cultivo INIA. National Institute of Agricultural Research.

[17] Murga-Orrillo, H., Coronado Jorge, M. F., Abanto-Rodríguez, C., & Lobo, F. D. A. (2021). Altitudinal gradient and its influence on the edaphoclimatic characteristics of tropical forests. Madera y Bosques, 27(3), e2732271, 1-13. https://doi.org/10.21829/myb.2021.2732271

[18] Boza, F. V. y Valery de Vélez, G. 1.990. Food plants of Venezuela: autochthonous and introduced: nomenclature, history, classification, cultivation, composition, utilisation and bibliography. Monograph number 37 (La Salle Natural Sciences Society). La Salle Natural Science Society.

[19] Barry, G, R., Rooney, T. P., Ventura, S. J. and Waller, D. 2001. Evaluation

of biodiversity value based on wildness A study of the western Northwoods Upper Great Lakes USA. Natural Areas Journal, 21(3), 299-242.

[20] Guerrero, R., Lugo, L., Marín, M., Beltrán, O., De Pinto, G. L. & Rincón, F. (2008). Physicochemical characterisation of the fruit and pseudofruit of Anacardium occidentale L. (merey) under dry conditions. Revista de la Facultad de Agronomía, 25(1), 81 - 94. URL:http://ve.scielo.org/scielo.php?script=sci_arttext&pid=S0378-78182008000100005&lng=es&tlng=es

[21] Cashew Production System (2016). Production System Data. Embrapa Agroindústria Tropical. Production System, 1. Electronic Version. (2nd edition). 1-193. https://www.spo.cnptia.embrapa.br/conteudo?p_p_id=conteudoportle...

[22] Lemos, Moaciria de Souza, Bordallo, Patricia do Nascimento, Vidal Neto, Francisco das Chagas, Lima, Eveline Nogueira, & Holanda, Ioná Santos Araújo. (2021). Genetic and physicochemical diversity in sweet Brazilian cultivars of anacard (Anacardium occidentale). Acta botanical mexicana, (128), e1775. https://doi.org/10.21829/abm128.2021.1775

[23] Borjas, M. (2015). Semi-industrial production of a jam from the pseudofruit of the caujil (Anacardium occidentale L.) (Master's thesis in Food Science and Technology). University of Zulia. Venezuela.

[24] Pérez, M. and Velázquez, F. Artisanal production of citric acid and vitamin C (ascorbic acid) from cashew nuts (Anacardium occidentale L.) and mangoes (Mangifera indica). (Special Degree Work in Chemical Engineering). Rafael Urdaneta University. Maracaibo, Venezuela (2021).

[25] Cardozo, R. (2022). Modelling the kinetics of osmotic dehydration and kiln drying of the Acadardium occidentales L. pseudofruit for raisin production in Zulia state. (Master's Thesis in Food Science and Technology). University of Zulia. Venezuela.

[26] Garruti, D. de S., Lima, J. R., Lima, A. C., Paiva, F. F. de A., Barros, M. E. S., & de Moraes Í. V. M., Pinto de Abreu, F. A , Machado, T. F., Rocha Bastos, M. do S., da Silva Neto, R. M., de Souza Filho, M. de S. M., & Nassu, R. T. (2015). Industrial Utilisation. In J. P. P., de Araújo (Ed.) Cashew (pp. 188-238).

[27] Sivagurunathan, P., Sivasankari, S., & Muthukkaruppan, S. (2010). Characterisation of cashew apple (Anacardium occidentale L.) fruits collected from Ariyalur District. Journal of biosciences research, 1(2), 101-107.

[28] Pereira, A. L. F., Maciel, T. C., & Rodrigues, S. (2011). Probiotic beverage from cashew apple juice fermented with Lactobacillus casei. Food Research International, 44(5), 1276-1283. https://doi.org/10.1016/j.foodres.2010.11.035

[29] Prommajak, T., Leksawasdi, N., & Rattanapanone, N. (2014). Biotechnological Valorisation of Cashew Apple: A Review. Chiang Mai

University Journal of Natural Sciences, 13(2), 159-182.

[30] Cruz Reina, L. J., Durán-Aranguren, D. D., Forero-Rojas, L. F., Tarapuez-Viveros, L. F., Durán-Sequeda, D., Carazzone, C., & Sierra, R. (2022). Chemical composition and bioactive compounds of cashew (Anacardium occidentale) apple juice and bagasse from Colombian varieties. Heliyon, 8(5), e09528.

[31] Quijada, O., Ramírez, R., Sindoni, M., Hidalgo, P., Mármol, E., Camacho, R., González,C. y Casanova, A. (2012). Preliminary behaviour of woolly clones precoces de merey (Anacardium occidentale L.) CCP-76 y CCP-1001 en la planicie de Maracaibo, Venezuela. Revista Científica UDO Agrícola, 12 (1), 25 - 31.

[32] Guerrero, R., C. Hernández, J. Chacín, C. Clamens, D. Pacheco, A. Sánchez- Urdaneta and B. Bracho (2014). Aspectos preliminares de la biología floral de Anacardium occidentale L. (Merey) en la altiplanicie de Maracaibo. Rev. Fac. Agron. (LUZ). 31(Supl. 1), 404-413.

[33] Baptista, A., R. Vilela, J. Bressan and M. Gouveia (2018). Antioxidant and antimicrobial activities of crude extracts and fractions of Cashew (Anacardium occidentale L.), Cajui (Anacardium microcarpum), and Pequi (Caryocar brasiliense C.): A systematic review. Oxid. Med. Cell. Longev. 139(1), 1-14.

[34] Peñalver, C. and Rodríguez, J. (2003). Food Standardisation. FAO-SENCAMER. National Seminaron CODEX Management. Strengthening the management ofthe national CODEX committeesin the Andean countries (Bolivia, Colombia, Ecuador,Peru,Venezuela),Project TCP/RLA/2904. Importancia del CODEX alimentarius enla food securityy el comercio de alimentos (FAO/RLC). Joint Programme FAO/WHO Programme on Food Standards since 1962 http://saber.ucv.ve/bitstream/10872/22016/1/Tema%2013.%20Normalizacion.%20Norm as%20ISO%2C%20HACCP.pdf

[35] Comisión Venezolana de Normas Industriales (COVENIN) (2015). Frutas Definiciones Generales (1ra. Revisión). 1834: 2015. FONDONORMA. Venezuela.

[36] Ewel, J., Madriz, A. y Tosi, J. (1976). Venezuela's life zones. Explanatory Memo on the Ecological Map. Ministry of Agriculture and Livestock.

[37]Comisión Venezolana de Normas Industriales (COVENIN) (1979). 1315-79. Foodstuffs. Determination of pH (ionic acidity). FONDONORMA.
[38] Comisión Venezolana de Normas Industriales (COVENIN) (1977). 1151-77. Fruits and derived products. Determination of acidity. FONDONORMA.
[39] Comisión Venezolana de Normas Industriales (COVENIN) (1983). 924-83: Frutas y productos derivados. Determination of soluble solids by refractometry (1st rev.). Technical Committee CT-10: Foodstuffs. Technical Subcommittee SC 6: Frutas y productos derivados en su reunión No. 22-06 de fecha 09-08-1983. Caracas. Venezuela. Fondonorma.
[40] Central American technical regulation 67.01.33:06. Processed food and beverage industry. Good Manufacturing Practices. General Principles. Adaptation of CAC/RCP-1-1969. Revised 4-2003. Recommended International Code of General Food Hygiene Practices. Available at https://asp.salud.gob.sv/regulacion/pdf/rtca/rtca_67_01_3306_bebidas_procesadas _buen as_practicas.pdf
[41] Codex Alimentarius international food standards, guidelines and codes of practice contribute to the safety, quality and fairness of this international food trade. Available at https://www.fao.org/fao-who-codexalimentarius/en/
[42] General standard for fruit juices and nectars. CXS 247-2005. Codex Alimentarius Commission Adopted in 2005. Amended in 2022.
[43] Coronado, M. & Hilario, R. (2001). Making marmalades. Food processing for small and micro agro-industrial companies. Editorial CIED.
[44]Barona, A. M. (2007). Mermeladas. Revisiones de la Ciencia, Tecnología e Ingeniería de los Alimentos, 7(1), 1-39.
[45] Codex Alimentarius International food standards (2020). Standard Ford jams, jellies and marmalades (CXS 296-2009).
[46] Comisión Venezolana de Normas Industriales (COVENIN) (1994). Norma Venezolana 2005:1994, Alimentos colados y picados, envasados para lactantes (1era Revisión). Available at: www.sencamer.gob.ve
[47] CODEX STAND 79-1981. Codex standard for jams (fruit preserves) and jaleas, Available at: www.codexalimentarius.org
[48] Comisión Venezolana de Normas Industriales (COVENIN) (2001). Norma Venezolana 2952:2001, Norma general para el rotulado de los alimentos envasados (1er Revisión). Available at: www.sencamer.gob.ve
[49] Statistical Analysis Software (SAS) Institute, Inc. (2000-2003). SAS User's Guide: Statistic. SAS Version 9.0. Institute, Inc., Cary, NC, USA.
[50] de Oliveira, V.H.(2002). Influence of irrigation on peduncle and nut production in early dwarf cashew clones. Revista Brasileira de Fruticultura, 24(3), 717-720.

[51] Paiva, J.R. and Cavalcanti, A. T. (2012). Cashew clones recommended by Embrapa Agroindústria Tropical. Fortaleza, CE: Embrapa Agroindústria Tropical. https://bit.ly/3JSGpME

[52] Avilán, L., Leal, F. and Bautista, D. (1989). Manual de Fruticultura. Merey. Editorial América. C. A. Caracas, Venezuela (pp. 415-441).

[53] Bezerra Almeida, M. L., Herbster Moura, C. F., Innecco, R. and Souza da Silveira, M. R. (2018). Physical characteristics of peduncles of dwarf cashew clones (Anacardium occidentale L.) produced as a function of environmental and temporal variation. Revista Colombiana de Ciencias Hortícolas, 12(1), 41- 49.

[54] Barrera, J., Arrazola, G. and Cayón, D. (2010). Physicochemical and physiological characterisation of the ripening process of Hartón plantain (Musa AAB Simmonds) in two production systems. Acta Agronómica, 59(1), 20 - 29.

[55] Aular, J., and W. Natale. 2013. Mineral nutrition and fruit quality of some tropical fruit trees: guava, mango, banana and papaya. Revista Brasileira de Fruticultura. 35(4):1214-1231.

MIX
Papier aus verantwortungsvollen Quellen
Paper from responsible sources
FSC® C105338

Printed by Books on Demand GmbH, Norderstedt / Germany